The Hurricane Hoax

and Other Cases

Seymour Simon
Illustrations by Kevin O'Malley

Volume in 5 the Einstein Anderson series

Published by Seymour Science LLC.

These stories, which have been substantially updated and expanded for a new audience, are based on the Einstein Anderson book originally published in 1981 by Viking Penguin, under the title "*Einstein Anderson Makes Up for Lost Time*" and republished in 1997 by Morrow Junior Books, New York, under the title "*Einstein Anderson: Science Detective, The Gigantic Ants and Other Cases.*"

www.StarWalkKids.com

ISBN: 978-1-936503-24-7 (Print)

Contents

The Hurricane Hoax

"Einstein, look at this. It's really scary."

Einstein Anderson's best friend, Paloma Fuentes, handed him her phone. He pushed his glasses back on his nose and looked at the screen. It was a video of ocean waves breaking over a house near the seashore. He nodded grimly.

"Yeah, that's pretty bad," he said, handing the phone back.

"I wonder if this Dr. Raynes really has invented something that can stop hurricanes," Paloma said. "Is it possible?"

"I guess that's what we're going to find out," Einstein replied. "If he could, it would be incredible. It would save thousands of lives and billions of dollars."

It was a Tuesday evening in mid-April and they were sitting in the middle of Sparta High School's auditorium. All around them people were filing in, looking for seats. But Einstein and Paloma were the only sixth graders there.

Some of the adults looked pretty worried. A hurricane had recently torn through the town of Sparta, blowing over trees, knocking down power lines, and causing a lot of damage.

Luckily no one had been hurt, but it had been very scary.

Now, someone named Dr. Raynes had called an emergency town meeting at the high school. He said he had a plan for dealing with future hurricanes. He'd even taken out an ad in the *Sparta Tribune* and placed notices on local blogs. Einstein's mother, Emily Anderson, was an editor and reporter for the *Tribune*. She decided to attend the meeting to see what Dr. Raynes had to say and she invited Einstein along, just in case she needed a science expert.

Anyone who knew Einstein wouldn't have been surprised that his mother sometimes turned to him for help with science facts. Even though he was only twelve, Einstein Anderson was famous in the town of Sparta for his amazing knowledge of science and for the way he used science to solve mysteries, both big and small. That's how he got the nickname, Einstein, after the greatest scientific genius of the twentieth century. His real name was Adam, but no one called him that anymore, not even his parents.

Albert Einstein was a world-famous thinker who came up with the equation $E=mc^2$. That simple equation helped lead to atomic energy and a new understanding of the universe. Einstein Anderson, however, was just an average-looking twelve-year-old kid with light brown hair and glasses that seemed too big for his face.

He and Paloma had been friends for a few years. They both went to Sparta Middle School. Paloma was the only person he'd met, or at least the only person his age, who loved science as much as he did—or maybe more. Paloma was taller than Einstein and she always wore her straight black hair in a ponytail, just like she always wore her red canvas high-top sneakers.

Emily Anderson turned to her son. "Einstein, I spoke with Dr. Raynes briefly. He says he has a way to stop hurricanes. I don't understand how anyone could claim to be able to stop hurricanes," she said. "I mean, some of these storms are hundreds of miles across. How could you stop that?"

"Well, some researchers have talked about it," Einstein told her. "Especially since hurricanes seem to be getting bigger and bigger, thanks to global warming. The way you would stop a hurricane is to do something about the heat."

"Heat?" his mom asked.

"Yes," Paloma explained, picking up where Einstein left off. "Hurricanes form over the ocean in the tropics, where the water is warmed by the sun. The air over the ocean heats up and, as you know, hot air rises."

"Yes, I did know that," Emily Anderson said with a smile. She was also used to having Paloma explain things to her.

"Well," Paloma continued, sounding a little bit like a professor, "the more heat in the ocean, the more the hot air rises. But other, cooler air has to come in to replace the hot air. Then that air heats up, and it rises. And if that keeps happening, you get a whirlpool of air rushing in—that's a hurricane."

"And that's the reason we're getting bigger hurricanes," Einstein added. "As the average temperature of the globe rises, the temperature of the oceans is going up. More heat means bigger and more frequent storms."

"Well, after hearing that, I certainly hope this Dr. Raynes has a solution," Mrs. Anderson said.

HURRICANE

"Did you ask him what he's a doctor of?" Einstein said to his mom.

"I did," Emily Anderson replied with a frown. "He avoided the question, but I plan to ask him again tonight."

"All this talk about hurricanes reminds me of something," Einstein began, but both Paloma and his mom quickly interrupted him.

"No jokes, Einstein," Paloma warned.

"Einstein, must you?" his mother asked.

But when it came to corny jokes, Einstein Anderson could not be stopped.

"How does a hurricane see where it's going?" he said with a chuckle.

"That's easy, Einstein," Paloma replied. "With its eye!" Even though Paloma knew the punch line, Einstein laughed anyway.

Just then, the crowd hushed as a tall, good-looking young man dressed in jeans and a black turtleneck sweater walked out onto the stage. He had thick, wavy black hair and a big, confident smile. He grabbed the microphone like a pop singer and began talking quickly and excitedly.

"Hurricanes!" he cried. "For centuries mankind has wondered how the destructive force of these terrible storms can be stopped. Now, for the first time we have the answer. My name is Dr. Phillip Raynes, and that's what I'm going to talk about tonight."

There was a rumble from the audience as everyone reacted to this news. But the audience quickly quieted down and listened, as Dr. Raynes paced back and forth across the stage. While he talked, photographs of hurricanes and their damage were projected on the screen behind him. With each image of destruction, he became more and more excited.

Finally, he paused, and then said in a dramatic voice, "As you know, the secret to the strength of hurricanes is the heat from the ocean!" Einstein and Paloma nodded in agreement. "That's also their weakness. We can stop hurricanes if we can cool down the water in the ocean."

"Yeah, but how are you going to do that?" Paloma muttered.

As if he had heard her, Dr. Raynes replied, "I know you're asking, 'How are we going to do that?' The answer is—with icebergs!"

The crowd reacted with a hum of talk as a video started playing on the screen behind him. It was an animated view of a giant iceberg being towed across the ocean into a hurricane. Dr. Raynes went on for a few more minutes. The more he talked, the more the audience rumbled. It seemed to Einstein that some people were excited about the idea of stopping hurricanes. But others were angry that they had come out to hear this crazy idea.

On stage, Raynes gave his closing pitch.

"Now, usually I would write a proposal for a research grant from the government," he said with a big, knowing smile. "But we all know how slow the government is."

Several people in the audience nodded and laughed.

Dr. Raynes nodded and continued. "That's why I've decided to build my hurricane halting machine privately—by forming my own company, Hurri-Can't, Incorporated. And you, lucky enough to be here today, can be among the first investors!"

Some people applauded, but others shook their heads. When things quieted down, Dr. Raynes looked out over the audience.

"Now, I'm sure some of you have questions," he said. "Who will be first?"

Almost before the words were out of his mouth, Paloma raised her hand. Dr. Raynes's face lit up with a big grin.

"Yes, young woman," he said with amusement. "What's your question?"

Paloma stood up.

"I'm Paloma Fuentes," she said. "And I don't see how you're going to get an iceberg big enough to cool off the ocean." Paloma wasn't very big, but her voice carried everywhere in the auditorium.

"Well, it's rather complicated, I'm afraid," Raynes replied. "Let's just say I don't need a really giant iceberg. You see, hurricanes are formed from high-pressure systems. The high pressure pushes the air outward in all directions. So the iceberg doesn't have to cool off the whole ocean, just disrupt the high-pressure air pattern. Did you understand that?"

"No," Paloma said with an angry frown.

"You could look at my website," Dr. Raynes said, very kindly. "It has a whole kids section that explains everything. Uh, next question?"

As Paloma sat down, she muttered, "I didn't understand it because it doesn't make any sense."

Now Einstein had his hand up. On the stage, Dr. Raynes laughed.

"My goodness," he said. "We have another young questioner. I'm glad that young people are so concerned about the environment. And what's your name, young man?"

Einstein stood up.

"Einstein Anderson," he answered, but his voice squeaked as he said it. A few people laughed. Dr. Raynes looked very serious.

"Einstein? Really?" he said. "Ladies and gentlemen, it seems we have a genius in the audience. Well, Einstein, what's your question? Do you also want proof that my machine will work?"

"Einstein is my nickname," Einstein said, very calmly. "And I don't have a question. I also have no idea if your machine will work, though I doubt it. But I can prove that you don't know anything about hurricanes."

Can you solve the mystery? How can Einstein prove Dr. Raynes doesn't understand hurricanes?

The smile on Dr. Raynes's face got even bigger.

"Really *Mr. Einstein*?" he said mockingly. "How will you prove that?"

Einstein pushed the glasses up on his nose. "I can prove it because what you said about hurricanes is exactly backwards," he replied. "You said that hurricanes are caused by high pressure systems. That's wrong. As the hot air rises at the center of a hurricane, it creates a big drop in air pressure. A hurricane is a large area of very *low* pressure, not *high* pressure at all. The low pressure is what causes the powerful winds to blow in a spiral toward the center."

As Einstein finished, Dr. Raynes's broad face slowly turned bright pink. For a moment, he was speechless.

"Young man, I'm...I'm sure you mean well," he sputtered. "But I think I know better than..."

"He's right!" someone shouted from the other side of the hall.

"Of course, I'm right," Raynes replied with a huff.

A man stood up near the back of the auditorium. He held up a smartphone.

"Not you!" he cried, then he pointed to Einstein. "Him! The kid is right! I just looked it up online."

The room erupted, with everyone talking at once. On either side, Einstein could see people taking out their phones and checking for themselves.

Paloma stood up and, in a clear voice that carried over the din, she shouted, "Of course he's right! That's why they call him *Einstein*!"

Raynes looked from left to right. Some folks in the audience had even started booing him. Without another word, he hurried from the stage.

"Well, this is going to be an interesting article," Mrs. Anderson said, as they left the auditorium. "Thanks to you, *Einstein*."

"I wonder what that guy was a doctor of," Paloma said. "Probably of fakeology."

"Hey, that reminds me!" Einstein said with a laugh.

Paloma groaned, "Oh, no, here it comes!"

But Einstein's mother nodded, "Go ahead Einstein, you earned it."

"Do you know why the house needed to a see a doctor?" he asked. Then before anyone could answer, he burst out with, "Because it had window pains!"

From: Einstein Anderson

To: Science Geeks

Experiment: How to Build Your Own Barometer.

My friend Paloma knew that Dr. Raynes was a fake the minute he said hurricanes were caused by wind circulating around an area of ***high*** pressure, because she knew that the centers of hurricanes are areas with very *low* pressure. But what's this air-pressure thing all about? How do changes in air pressure affect our weather? And how do meteorologists (that's people who study the weather—like the weatherperson on TV) measure air pressure?

Let's do a demonstration to show how powerful air pressure is. Then we can build our own barometer to measure air pressure and start to predict the weather!

Here is what you need to see air pressure in action:

- A wide-mouth drinking glass
- A square of glossy smooth cardboard a little larger than the mouth of the glass

- Water
- A scale

Note: Do this over a sink until you get the hang of it.

Weigh the empty glass on your scale. Fill the glass to the top with water and weigh it again. Make a note of the weight of the water (subtract the weight of the glass from the total of glass + water).

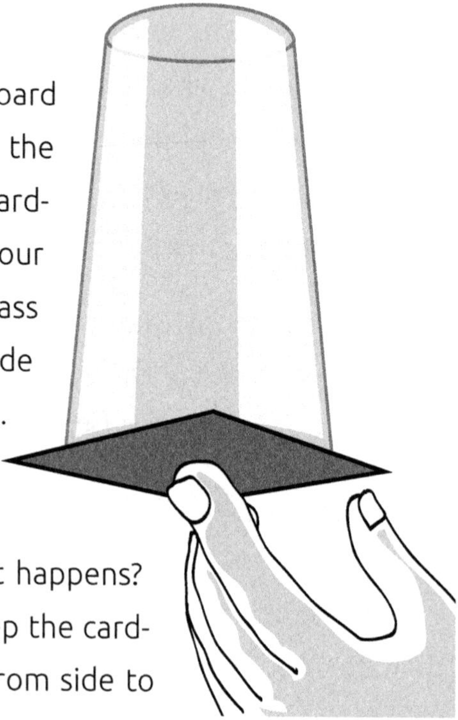

Now put the cardboard over the mouth of the glass. Holding the cardboard in place with your hand, turn the glass over so that it's upside down over the sink. Carefully remove your hand from the cardboard. What happens? Using a finger to keep the cardboard from sliding from side to side, turn the glass sideways in all directions. What happens now? Does it make a difference which way you turn the glass? What do you think would happen if you pried back the cardboard and let air into the glass?

What's going on here?

Air doesn't weigh very much, but there is a lot of it in the atmosphere and Earth's gravity pulls it toward the ground in the same way that gravity keeps us from floating away. When you filled the glass with water, you got rid of all the air in the glass, and replaced it with water.

At sea level, the average air pressure is 14.7 pounds per square inch (1.03 kilograms per centimeter). What kept the card in place was the force of that air pressure, pushing against the cardboard. There was no air in the glass so nothing to push back against the air outside. If you figure out the number of square inches covered by the cardboard and multiply that by 14.7, you will see that the air pressure exerts a lot more force than the weight of the water in your glass.

And you thought air was nothing!

So now, let's make that barometer.

Here is what you need:

- A ruler
- Modeling clay
- Water
- A bowl
- Clear plastic bottle
- String
- Paper
- Pen or pencil

Put a lump of clay in the bottom of the bowl and stick the ruler into it so it stands up straight. Pour enough water into the bowl to fill it about $^{1}/_{3}$ full. Then fill the clear bottle about $^{3}/_{4}$ full of water. Put your hand over the top to keep the water from running out and turn it over, upside down, into the bowl. Once the top of the bottle is under water, you can take your hand away and the water won't run out. Use your string to tie the bottle to the ruler.

Cut out a strip of paper 4 inches long and mark lines on it every ¼ inch. Use a dark line to mark the middle of the strip. Tape or glue this to the water bottle so that the dark line matches the level of water in the bottle.

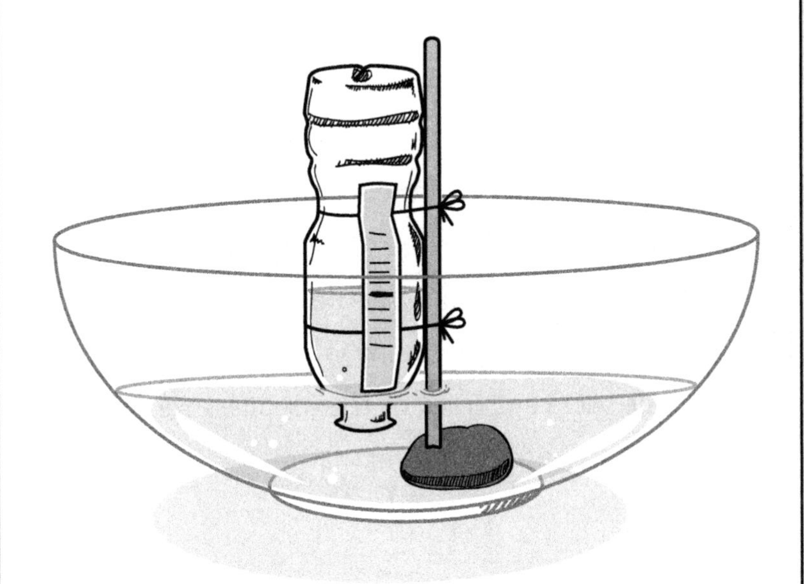

That's it! You have a water barometer. When the air pressure rises, it will press down with more force on the water in the bowl and force water up into the bottle. You will see that the water level goes up to a higher pressure line. When the air pressure falls, it will press with less force on the water in the bowl and some of the water in the bottle will flow out into the bowl. On a low pressure day, the water line in the bottle will fall below the dark line on your paper.

Look up your local weather online or on TV and keep track of the air pressure, or barometric pressure, for a few days. Your water barometer won't be as accurate as the ones the weather service uses, but you should see the water level change up and down along with the weather report. In general, high pressure goes with sunshine and good weather, while low pressure usually means cloudy or rainy weather. Did you notice that while you kept track of the air pressure with your water barometer?

The Science Solution:

So now, what does air pressure have to do with the weather? I'm glad you asked.

Let's keep in mind a few things: 1) Hot air rises. 2) Higher air pressure creates greater force and pushes things around. 3) Low pressure does not "pull" objects or particles, but when hot air rises it creates a vacuum, which heavier, colder air flows in to fill. The difference in barometric pressure between high pressure and low pressure areas causes wind—and hurricanes.

As Paloma said, hurricanes form in the late summer when water temperatures in the ocean rise above 80 °Fahrenheit (26.6 °Celsius). The warm water heats the air and makes water vapor rise into the atmosphere, where it cools and forms clouds. Colder air, in turn, comes in below and is warmed by the water until it rises. All this movement creates a low pressure area, ripe for conversion to a serious storm. If you are interested in hurricanes, check out this link: www.miamisci.org/hurricane/ to see a video with lots of details.

What is the highest air pressure reading you've had on your barometer? The lowest? Share your results with other Science Geeks at www.seymourscience.com.

2. The Snake Attack

CRASH!

The sound of something clattering in the floor woke Einstein Anderson out of a deep sleep. He looked around his room groggily, but it was dark. Then it happened again.

CRASH!

Einstein bolted upright in bed. The sound was coming from his little brother's room, next door. He grabbed his glasses from the bedside table and put them on, then swung his feet over and got out of bed. The floor was cold against his bare feet. It had turned very chilly overnight.

He went to the door, turned down the hallway. There was light coming from under Dennis's door. Before he even opened the door, Einstein knew what was making the noise.

"Stop dropping the fishing tackle," he said as he pushed open Dennis's door. "I'm trying to sleep."

His brother stood in the middle of the room, completely dressed and ready to go fishing. He had on long pants, high rubber boots, a sweatshirt, a fishing vest, and a baseball cap jammed over his brown hair. In one hand he held a fishing rod and a net, and in the other hand he held a fishing tackle box.

"What do you mean, sleep?" Dennis asked, his freckled face beaming with happiness. "It's time to go fishing!"

It was true. Their dad had promised to take them fishing up at Carter Lake that morning.

"It's not time yet," Einstein protested. "It's cold and I want to go back to bed."

"That's not true," Dennis said. "You're not cold. I heard you say so."

Einstein pushed his glasses up on his nose and looked at his brother.

"What do you mean, I said so?" Einstein asked.

"You said people are warm-blooded," Dennis replied. "So you can't be cold."

"I said that people are..." Einstein stopped in mid-sentence. He was a little shocked to find that Dennis had actually listened to him. "That's not what that means," he began again.

"Warm-blooded means your body creates heat. It doesn't mean you never feel cold."

"Now, *Einstein*," Dennis lectured him. "That doesn't make sense. How can you be cold if your blood is warm?"

Einstein sat on the edge of Dennis's bed. "Look. Cold-blooded animals, like most reptiles and insects, don't produce heat. They have to get heat from the sun or some other source. If it gets too cold, they slow down or hibernate."

"Bears aren't insects," Dennis objected.

"What do bears have to do with it?" Einstein replied, feeling very tired.

"You said insects hibernate. But bears hibernate and they're not insects."

"Yes, but..." Einstein scratched his head. Sometimes he thought that Dennis got science mixed up on purpose, just to annoy him. "Look, forget about hibernation. Cold-blooded animals don't make enough heat in their bodies to keep moving when the air is cold. Warm-blooded animals, like us, can keep active even when the air is cold."

"That's what I said," Dennis insisted. "You can't be cold because you're warm-blooded."

For once, Einstein felt confused. He hated being woken up early. "Forget it," he grumbled. "I'll go get dressed."

After a quick breakfast with their dad, and after making some sandwiches for lunch, the three of them packed their gear in the truck and began driving to Carter Lake. It was still cold, so they all had to wear warm jackets and hats.

"I'll have to leave you boys for a little while," Matt Anderson explained as he drove. "I have to go over to the Joneses' farm and check on his mare." Dr. Anderson was a veterinarian and some of his patients were animals on the farms outside of Sparta. "Is Paloma going to be at her aunt's house today?"

Paloma's Aunt Camilla had a house just off the lake.

"No," Einstein replied. "She stayed home."

"Well, Einstein," his dad said. "You'll have to keep an eye on your brother while I go to the farm."

"I don't want his eye on me!" Dennis cried. "That's yucky!"

Matt Anderson laughed. "Okay, he'll just look at you every now and then. How's that?"

"Hey, Dad. Talking about the farm reminded me of something…"

Matt Anderson and Dennis both groaned because they knew what was coming.

"Did you hear that Mrs. Norton's hens stopped laying eggs?"

"No, I didn't hear that," his dad replied, good-naturedly.

"Yeah," said Einstein. "Her hens are tired of working for chicken feed."

It took about a half hour to get to Carter Lake. It was still early and there weren't many cars in the parking lot.

"If we're lucky, we might catch some trout today," Dr. Anderson said as they unloaded their gear.

"And if we're not lucky, we'll catch a cold," Einstein added, stamping his feet.

"Come on!" Dennis shouted and ran toward the lake.

The three of them fished happily for about an hour, moving around the lake as they searched for a good spot. Dr. Anderson caught a smallmouth bass and Einstein caught a yellow perch, but he threw it back. Dennis grumbled and complained about not catching anything, but he seemed pretty happy to be there. Then it was time for their dad to visit the Joneses' farm.

"Stay around here," he told them. "Here's the pack with your lunches. I'll put it on this boulder, so you'll know where it is. Do you have your cell phone?" he added, looking at Einstein.

"Yeah," Einstein nodded. "And I have service. Plus, we can always go up to Aunt Camilla's house if we need anything. It's just over the hill."

"Okay," Matt Anderson replied. "I'll be back in about an hour." He said good-bye and walked away toward the parking lot.

"Okay," Einstein said to his brother. "How about we try another spot? I think just down there, where the trees hang over the bank."

Dennis looked where Einstein was pointing. "You mean where Stanley is taking a photo of the lake?"

"What?" Einstein pushed his glasses back on his nose. Sure enough, he saw a tall, thin, blond-haired kid walking along the shore, bundled in a blue, puffy down jacket. It was Stanley.

"What's he doing here?" Einstein wondered aloud.

Stanley Roberts was also in the sixth grade at Sparta Middle School. Einstein had known him since kindergarten. They didn't always get along, mainly because Stanley felt that he, not Einstein, was the real science genius in Sparta. The only problem was that Stanley didn't spend much time learning about science. Instead, he was always dreaming up get-rich-quick schemes. He had lots of ideas for inventions that would make him a billionaire, like Mark Zuckerberg of Facebook or Steve Jobs of Apple. He had even made up his own company, StanTastic Industries.

"I guess we'd better find out what he's up to," Einstein said to Dennis. "Come on."

The boys carried their poles and nets and walked along the lake. Stanley didn't see them at first. He was holding a small camera in one hand and a smartphone in the other.

"Hey, Stanley," Einstein said, trying to sound as friendly as he could.

Stanley was so startled that he nearly jumped.

"Oh!" he grumbled. "If it isn't *Einstein Anderson.*" Stanley liked to make fun of Einstein's nickname, although everyone knew he was just jealous. Once he had even tried to get people to call *him* Einstein.

"What are you two doing here?" he asked suspiciously.

"We're fishing," Dennis told him, holding up his fishing pole. "What do you think?"

"Well, I'm not doing anything interesting!" Stanley added quickly, trying to hide his camera and phone.

"Don't worry," Einstein said with a smile. "We're not going to try to steal your new invention."

"How did you know it was a new invention?" Stanley replied angrily.

"Because you always have a new invention," Einstein told him.

"Well," Stanley said with a huff, "It's a new app for smartphones. It will help you find fish."

"Really?" Einstein said. He was impressed. "I've heard about apps like that. Does yours use sonar?"

"Sonar?" Stanley looked confused.

"Well, how else are you going to find the fish?" Einstein asked.

Stanley opened his mouth, but Einstein knew he didn't have an answer. Luckily for him, Dennis interrupted. "Einstein, I'm hungry," he complained. "Let's eat lunch."

"Now? I thought we'd wait for Dad," Einstein replied. "Let's try this new spot first, for just a few minutes."

"All right," Dennis agreed. "But if we catch a fish, I might eat it on the spot."

"No need for that," Einstein laughed. "The sandwiches are just over there, on that rock."

"Sandwiches?" Stanley repeated. He looked in the direction Einstein had pointed.

"*Our* sandwiches!" Dennis said sternly.

"Yes, *your* sandwiches," Stanley replied. "Big deal. I can get sandwiches anytime.

Now, if you'll excuse me, I have to test my app. I mean, my app with *sonar*."

Einstein and Dennis moved down toward the spot where the trees hung over the lake. Meanwhile, Stanley moved in the other direction, toward the parking lot. The two brothers got their lines in the water and quickly forgot about Stanley as they patiently waited for the fish to bite. But after only ten minutes, Dennis began to complain.

"Come on, Einstein," he whined. "I'm hungry, so let's go eat. Dad won't mind."

"Okay, I give up," Einstein said with a smile. "If the fish aren't going to eat, at least we can."

They turned to walk back along the lake when, just at that moment, they heard a loud yell. It was Stanley. They could see he was jumping up and down, right next to the boulder where they had left their sandwiches and other equipment. He was shouting. The two boys dropped their fishing gear and ran toward him.

Einstein got there first.

"What? What's wrong?" he asked, looking around. He already had his cell phone out. "Do you want me to call for help?"

Stanley was suddenly very quiet.

"Help?" he said, sounding insulted. "Why would I need help?"

"Because you were screaming," Dennis said as he ran up.

"Oh, that..." Stanley said. "I was just..."

But Dennis interrupted."Hey!" he shouted. "Where are our sandwiches?"

Einstein looked around. It was true. The daypack with the sandwiches was gone.

"They must be around here somewhere," he said. "Stanley, why were you shouting?"

"I . . . uh, I," he stammered. "I was just testing my app," he blurted out. "Testing the sonar."

"You can't test sonar by shouting," Einstein told him.

"Uh, yes, that's what I discovered," Stanley replied.

Meanwhile, Dennis had been busily searching for the daypack.

"There it is!" he cried from the other side of the rock. "It fell into a hole. I'll get it."

"No!" Stanley screamed. "Don't!"

Einstein and Dennis looked at Stanley.

"What's wrong with you?" Dennis asked.

"Don't reach in that hole," Stanley said, looking very nervous. "There's a snake down there."

Dennis jumped back. "A snake?"

"Yes!" Stanley replied, looking very serious. "I saw it slither in there."

"That's ridiculous," Einstein told them as he

walked around the boulder. "Watch out, Dennis, I'll get the pack."

"I'm warning you, Einstein," Stanley said urgently. "Don't reach in there unless you want to get bit by a poisonous snake."

"A poisonous snake?" Einstein repeated.

"That's right!" Stanley insisted. "I saw it when . . . when I was testing my app."

Dennis looked very worried. "Maybe he's right, Einstein. Don't reach in there. Forget the lunch. We'll just wait for dad."

"Well, I never believed you'd say wait for lunch," Einstein said with a chuckle. "But there's no need. I'm sure Stanley didn't see a snake slither in there."

Can you solve the mystery? Why was Einstein sure there was no snake in the hole?

As Dennis and Stanley watched, Einstein reached down into the hole. Suddenly, a look of pain shot over his face. His hand seemed to be stuck. Then he yanked it back and a long, thin, black object flew through the air and landed at Stanley's feet.

Stanley screamed loudly and jumped in the air.

"*A snake!*" Dennis cried in alarm. Then he started to laugh. "Hey!" he shouted. "That's no snake. That's a stick!"

Stanley looked at the ground near his feet. There lay a thin, black stick.

"What a lousy trick!" he complained.

"How'd you know it wasn't a snake, Einstein?" Dennis asked. "I'd have been too scared to touch it."

"Stanley said he saw the snake slither past. But it's too cold outside for a snake to be moving around. That's because..."

"Snakes have cold blood!" Dennis shouted.

"They're cold-blooded," Einstein said with a smile. "But close enough. So I knew a snake couldn't be active today. Plus, there are very few poisonous snakes in North America."

"It was still a lousy trick," Stanley repeated.

"Not as lousy as trying to sneak one of our sandwiches," Einstein told him.

"I wasn't trying to sneak one," Stanley whined.

"Then what were you doing with our day-pack?" Einstein said. "It didn't fall off that rock by itself."

"I was just..." he began, but then his shoulders slumped. "I just wanted to see what kind you had," he said. "I forgot mine back home.

Then I saw that snake—I mean that stick—and I guess I freaked."

"That's 'cause you had a guilty conscience!" Dennis told him crossly.

"You know what?" Einstein said to his brother. "Stanley's had a rough time. I think we should share our lunch with him. Dad made a couple of extra, anyway."

"Okay, Einstein," Dennis grunted. "As long as we eat now! Snakes make me hungry."

"Thanks, guys," Stanley said, looking very relieved. "As a reward I'll let you invest in my fish finding app!"

"No thanks," Einstein said. "But that reminds me, do you know what kind of fish likes it when the lake freezes?—A skate!"

From: Einstein Anderson

To: Science Geeks

Experiment: Chill Out: Make a Frog Hibernate

(If you don't have access to a frog, you can do this with an insect.)

Stanley will do anything to get what he wants, whether it's a million dollars or a PB&J sandwich, except learn the science! Even my little brother, Dennis, is onto him now. But his little trick with the "snake" that turned out to be a stick made me think about some cool facts about cold-blooded animals.

What would it be like to be cold-blooded? Let's do an experiment and see.

Here is what you need:

- A frog (or insect)
- A large-mouth gallon jug (or deep plastic drinking glass for the insect)
- Mesh to put over the top of the jug or glass to keep your animal from taking a hike.

- A shallow metal pan
- Ice
- Salt
- Water
- Laboratory thermometer

If you live in a place with ponds and woods, you may be able to find a frog for this experiment. Look around the edges of the pond or, if it's fall and the weather is cooling, look for frogs beginning to hibernate by bedding down in the mud under the water.

Put your frog in the glass jar with a few inches of water in the bottom and cover the top with a screen. Insert your thermometer through the screen into the water. Do not add water to the glass if you are using an insect. Write down the temperature of the water (or the air temperature in the glass, as measured by the thermometer). Count the number of breathing movements per minute by watching the underside of the frog's mouth. Note if the frog moves his eyes or legs. Write down what you observe about the insect's movements.

Now, place the jar (or glass) in the pan and pack ice around it. Add some salt to the ice to make it even colder. Watch the thermometer and repeat your observation of the frog's breathing and movements after

the temperature has dropped five degrees. Keep watching and noting changes in the frog's behavior every 5 degrees. Once it begins to slow down, write down your observations every 2 minutes. You can make a graph that tracks the number of breathing movements per minute at every 5 degree interval. You should see the numbers getting smaller and smaller. If you are working with an insect, you should notice the insect's movements slowing down.

Finally, when the temperature is cold enough, the frog, (or insect) will begin to hibernate. You will recognize the beginning of hibernation by these behaviors:

- The frog's eyes will become covered by a membrane.
- It will blow air from its lungs and dive down to the bottom of the jar, staying under water.
- It may try to "dig" on the bottom of the jar before it settles down. In the natural environment, the frog's digging would have buried it in the protective mud at the bottom of a pond. It will begin to breathe by absorbing oxygen from the water through its skin.

Once your frog is hibernating, remove the jar from the pan of ice. Continue to observe the frog at intervals as it warms up, counting its breaths and seeing it gradually become more active.

If you are doing this experiment with an insect, you will need to lower the temperature by about 30 °F (16 °C). You may want to put the insect in its cup into the refrigerator to get this temperature difference. But do not put it into the freezer! As with the frog, observe the insect's movements and how they change, as the temperature gets colder. When the insect appears to be hibernating, return it to a terrarium or its natural environment and observe what happens as it warms up.

The Science Solution

Most mammals and birds are warm-blooded, meaning that they maintain a steady temperature, regardless of the temperature outside their bodies. Reptiles, insects, arachnids (spiders), amphibians, and fish are cold-blooded, meaning that their body temperature changes with the exterior temperature.

Mammals stay warm by converting the food they eat into energy, and they use various strategies to cool down: primates like humans and apes have sweat glands all over their bodies to help them stay cool, but dogs and cats only have sweat glands in their feet. Large mammals, like elephants, have a harder time cooling down if they get overheated. That's why elephants have such big, thin ears—the ears lose heat quickly and cool the elephant's brain.

Believe it or not, there are some advantages to being cold-blooded. For example, cold-blooded (also called ectothermic) animals don't have to eat as much as we warm-blooded (or endothermic) creatures do, because they use much less energy keeping themselves warm. So in an environment where food is scarce, like a desert, reptiles have an advantage over mammals. Also, since reptiles' bodies are cooler than mammals', they don't offer such an attractive environment for bacteria and viruses that make us humans sick.

As you noticed in this experiment, cold-blooded animals are more active when the temperature is warmer and are very sluggish when it gets cold. This is because their muscle activity depends on chemical reactions that run quickly when it is warm and slower when it is cooler. As we also saw, when animals hibernate their breathing and heart rates slow way down. Some small mammals like hamsters can hibernate. When hamsters hibernate their heart rate goes from 400 beats a minute to about four or five!

I need a snack to keep me warm!

3. A Cold Case

A few days after they went fishing at Carter Lake, Einstein went over to Paloma's house after school. They had a team project due for science class.

Paloma lived in a large, old house, painted white with green trim and set back on a wide lawn. Einstein climbed the steps to the front door and knocked. Paloma's mother, Adrianna, opened it. Like Paloma, she was tall and thin and had a long ponytail of black hair, except hers was beginning to have gray in it.

"Well, if it isn't Mr. Einstein," she said with a big smile. "Have you come to blow up the house again?"

"Hi, Mrs. Fuentes," Einstein replied politely. "No, we're just going to build a nuclear reactor in the attic, if that's okay. I promise it won't blow up."

Mrs. Fuentes made a show of thinking it over.

"Well, I suppose that's all right," she said, opening the door wider to let Einstein in. "As long as you clean up when you're done."

"We'll sweep up all the uranium," Einstein told her.

"Paloma's upstairs in her laboratory," Mrs. Fuentes said, pointing up the wide staircase.

Einstein nodded and went up the stairs, two at a time.

Paloma's "laboratory" was really a sunroom on the second floor, next to her bedroom. Her father had set up a plain wooden workbench in the sunroom, and that's where Paloma built her robots, tinkered with computer parts, and kept her aquarium and terrarium. When Einstein walked in, she was just feeding her tropical fish. They were all at the surface of the water, grabbing the food with wide-open mouths.

"Hey, *Einstein*," Paloma called out. Like her mother, Paloma liked to tease him about his nickname, but Einstein didn't mind. He knew it was kind of a funny nickname for a twelve-year-old.

When she was in her lab, Paloma wore a white lab coat she had gotten from her Aunt Camilla. Paloma always said, "If you're going to be a scientist, you should dress like a scientist." And on her feet, of course, were her red canvas high-top sneakers.

Einstein looked at the fish in the tank and then turned to Paloma. "Do you know why the fish crossed the ocean?" he asked.

Paloma frowned. "No, *Einstein*" she replied. "I don't know why the fish crossed the ocean."

Einstein was already laughing at his own joke.

"To get to the other tide! Get it? Tide!"

"I get it," Paloma grumbled.

"Are you ready to work on the project?" Einstein said when he had finished laughing.

"Sure," Paloma answered. "I have the chromatography materials—the paper towels, the marker pens, the water. But before we get started, I promised my mom I'd put some stuff away."

"Well, I don't want your mom getting angry at me," Einstein said. "She already thinks I'm going to build an atom bomb up here."

"That's one thing you and my mom have in common," Paloma said, as she led him into her room.

"Our interest in atomic energy?" Einstein asked.

"No, your interest in dumb jokes!"

Einstein saw that Paloma had piled up all her winter gear—ice skates, cross-country skis, warm socks, and wool hat.

"I have to find a place for all this in my closet," she said, as she started to gather stuff up off the floor. Einstein picked up the skates.

"You know, I never understood why these are called ice skates," he said. "They should be called *water* skates. The pressure of your weight on the blades raises the temperature of the ice and melts it."

"I know," Paloma told him, as she stuffed the skis in the back of her closet. "You don't really skate on the ice, you skate on a thin film of water under the blades."

Einstein nodded. "So, put away your water skates and let's get to work."

The two of them went back to the lab and began their chromatography experiment. They drew on the paper towels with different pens, then placed the edge of the towel in water. As the water crept up the towel and crossed the line of ink, different dyes from the ink separated out. This made a rainbow pattern on the paper towel. Each pen left a different rainbow "fingerprint."

Einstein and Paloma took turns recording the results with a video camera. Later, they would upload the video to their class YouTube account.

"You could use this sort of experiment to figure which brand of pen was used to write a

note," Einstein pointed out. "It would be a good way to solve a mystery."

"That's true," Paloma replied with a laugh. "Except it doesn't work on e-mail messages."

"Hey, *Einstein,*" she continued. "I just had a brainstorm!"

"Uh-oh, I'd better get my umbrella," Einstein said.

"Very funny," Paloma replied. "But you won't be laughing when I finally stump you! All that talk about ice gave me a great idea. I have a challenge for you that you won't be able to solve."

Einstein pushed his glasses back on his nose. Every now and then, Paloma liked to challenge him with some science mystery or riddle. She hadn't stumped him yet, but she kept trying.

"Okay," Einstein nodded. "Go ahead and try."

Paloma grinned. "Just wait here and I'll set the challenge up in the kitchen. And be prepared to admit defeat."

After about ten minutes, Paloma called up the stairs.

"Okay, *Einstein!*" she shouted. "Come downstairs and get stumped!"

Einstein hurried down to the large, brightly decorated kitchen. When Einstein walked in, Paloma opened the refrigerator and took out a large block of ice from the freezer compartment. She placed the ice block on top of a plastic dish rack in the sink.

"Einstein," she said, "here's the challenge. Can you cut through this chunk of ice and still leave one solid block of ice behind?"

"What?" Einstein asked. "How can you cut through ice and not get more than one piece?"

"That's for me to know and you to find out," said Paloma.

Einstein went over to the ice. "If I heated a knife or a piece of wire, it would cut through the ice," he said, thinking out loud.

"But then you'd have two pieces of ice left," said Paloma. "Do you give up?"

"Just a minute," Einstein said. He pushed back his glasses and was quiet. Then his face lit up. "If you give me a piece of wire and two bricks," he said, "I'll do it."

Can you solve the mystery? How can Einstein cut through a block of ice and still leave a solid block behind?

Paloma frowned. "Einstein!" she said, "Don't tell me you figured it out already!"

Einstein just shrugged. He knew Paloma was just dying to stump him.

"Sorry," he said. "Let's try it together."

"I don't have bricks," Paloma told him. "What about some brass weights?"

Einstein and Paloma tied a brass weight to each end of a piece of thin wire. Then they placed the weighted wire on top of the ice. Slowly the wire began to pass through the ice, yet it left a solid block behind.

"I can see why talking about ice skates made you think of this," Einstein said while they watched. "It's the same idea. The ice directly under the wire melts because the pressure of the wire lowers the melting point of the ice. The ice changes to water, and the wire slips down. But the water freezes again as the wire passes through and the pressure is released."

"Well, I guess you did it again, *Einstein*," Paloma grumbled. "But I'll get you next time."

"We'll see," Einstein said. Then he began to snicker. "By the way, do you know why you should never tell a joke while you're ice skating?"

He looked so happy that Paloma smiled in spite of herself. "No, Einstein," she said, laughing. "Why?"

"Because the ice might crack up!"

From: Einstein Anderson

To: Science Geeks

Experiment: Crystals You Can Keep.

That experiment Paloma did with the block of ice and wire was really cool. You can do it at home and see for yourself that pressure melts ice. But watching the ice made me think about the beautiful crystals that form when water freezes (think snowflakes) and I started to wonder if other substances, besides water, make crystals.

Let's freeze some water to make water crystals and then do an experiment that will give us crystals that won't melt!

Here is what you need:

- 2 plastic food storage containers with lids
- A stove or hot plate and pan to boil water (You will need an adult to help you with this.)
- A freezer
- A magnifying glass

- Epsom salts
- Liquid glue
- A piece of window glass
- Newspaper
- Paper towels

Fill one of the food containers about halfway with water, then put on the lid and shake it to get air into the water. Put this container into the freezer. Then ask your adult helper to boil some water for you and carefully pour the boiling water into the other food container and put the lid on, trapping the steam inside and being careful not to shake up the hot water. Put this container into the freezer too.

The next day, open your two containers and compare what you see. In the container with hot water, the steam should have formed crystals on the inside of the lid. Look at them with a magnifying glass. Can you draw what you see? Do the crystals form a pattern?

Now, remove the blocks of ice from both containers by running warm water over the outside of the containers. Do you notice any differences between the two ice blocks? The one you shook up should be cloudy in the center. If you look closely you may see worm-holes made by the air trapped in the freezing water. You can use these blocks to repeat the experiment that Paloma and I did, using pressure to melt the ice and watching it glue itself back together again.

Now let's make some crystals we can keep:

Using the pan and hot plate, ask your adult helper to boil about 1/2 cup of water. After the water boils, remove the pan from the heat and add 1/3 cup of Epsom salts to the water. Add 3 or 4 drops of glue and stir until the Epsom salts are dissolved.

Carefully clean a piece of window glass with detergent and water. Rinse it well, dry it, and lay it down on some sheets of newspaper. Use a paper towel to spread the Epsom salts mixture on the glass. Wait a few minutes.

As the water evaporates, you will see long, needle-like crystals growing on the glass in all directions. Compare these crystals to the water crystals you drew before. Are they the same or different?

When the glass is perfectly dry, you can save your crystals.

The Science Solution

You probably know that matter can exist in three states: solid, liquid and gas. Different kinds of matter freeze/melt or turn to gas/boil at different temperatures. For example water freezes/becomes solid at 32°F (0°C) and boils/becomes a gas at 212°F (100°C) when it's at sea level. Boiling water converts the liquid water into a gas called water vapor. And, as we saw in our experiment, when you freeze water vapor, it crystallizes directly.

But why do crystals form? And why are crystals from different kinds of matter, like water and Epsom salts, different? The shape of a crystal depends upon the way the atoms are arranged to form molecules. The chemical recipe for water is H_2O, which means that a water molecule is made of two atoms of hydrogen and one atom of oxygen. The water molecule looks a little like a Mickey Mouse head, with oxygen for the face and the two hydrogen atoms attached to the oxygen at a 120° angle.

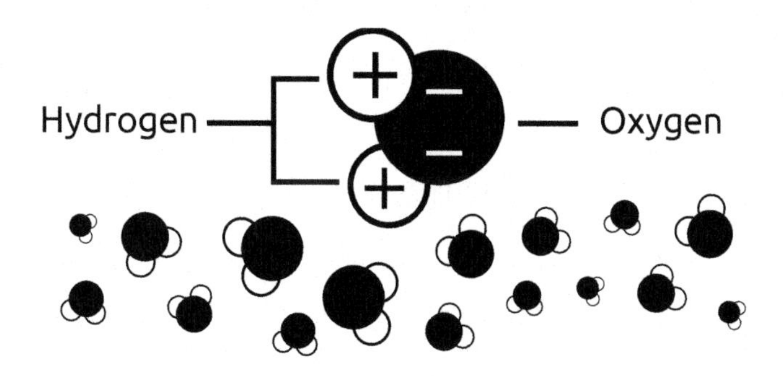

Water

When water becomes solid as ice, the more negatively charged oxygen atoms are attracted to the hydrogen atoms, so the water molecules arrange themselves in hexagons. Just like snowflakes!

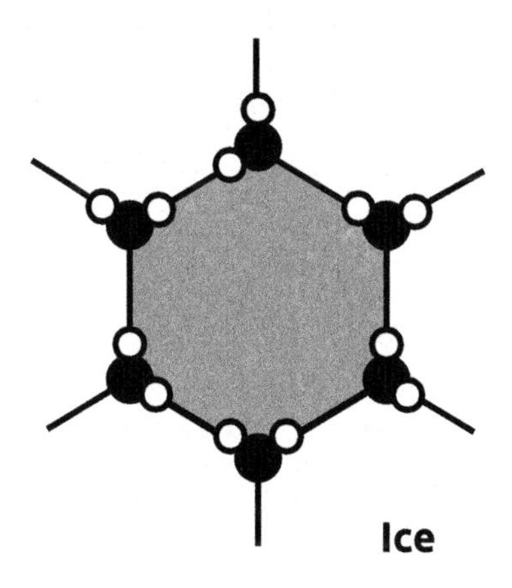

Ice

And why did the ice with air in it have worm-holes in the center of the block? Because ice freezes from the surface inwards. If there are impurities in the water, they won't fit into the hexagonal structure and they get pushed away, in front of the advancing ice crystals. The air got trapped in the center of the block because it could not escape through the frozen part of the ice, and it could not fit into the crystal pattern

By the way—this is very good news for the frog in our last experiment. If ice froze from the bottom up, fish and amphibians that hibernate in ponds would be frozen and they could not survive the winter!

If you are interested, you can make crystals with lots of different salts. Some that you may be able to find at your local pharmacy include: alum (potassium aluminum sulfate) or Rochelle salt (potassium sodium tartrate).

4. Under Pressure

"We felt a lot of pressure when we were working on this project. A lot of high pressure. It was a real pressurized situation."

Pat Hong stood in front of the class and waited for a laugh that didn't come. He and Jamal Henry were giving a report to Ms. Taylor's science class. The report was about their idea for a new sort of pressure cooker. It had seemed like a good idea to put in a joke to start things off, but now Pat was regretting it.

From his seat near the back of the class, Einstein Anderson could see Pat start to sweat.

"Uh-oh," he thought. "The only time Pat gets this nervous is when he tries to talk to Paloma."

As if she had read his thoughts, Paloma turned in her seat and shot Einstein a worried look.

Pat and Einstein had become friends in the summer between fifth and sixth grade. Pat was Sparta Middle School's star athlete, and he wasn't that interested in science. Einstein liked to play sports, but his real passion was for science. In spite of their differences, they had become good friends after Einstein solved the mystery of Pat's howling hound dog.

Pat stood at the front of the class. He'd only begun his talk, but now he wished it was already over. The assignment was to come up with an invention to present at the spring school fair. Each class got to present one invention that showed a practical use for science. They didn't have to build it, but they had to create diagrams and models that proved it would work.

Jamal and Pat had volunteered to come up with an idea for Ms. Taylor's class. Einstein and Paloma had offered to be on their team, but they said they'd rather do it themselves.

"That's…that's a joke," Pat stammered in front of the silent classroom. "Because our invention is a pressure cooker. Pressure—get it?"

Jamal quickly stepped over to a computer on Ms. Taylor's desk and hit a key. A colorful diagram appeared on the screen at the front of the room.

"Our invention is a new kind of pressure cooker," Jamal said hurriedly. Einstein noticed that Pat looked very relieved.

Jamal spoke confidently about their idea. "Pressure cookers work by raising the boiling point of water," he explained, as he clicked to the next slide. "They're tightly sealed, so any liquid that boils and turns to steam is trapped in the pot. That raises the pressure."

Jamal looked at his teammate. Einstein guessed it was Pat's turn to speak, and the tall athlete seemed to get over his nervousness. He easily picked up where Jamal had left off.

"Under normal pressure, the boiling point of water is 100 degrees Celsius, or 212 degrees Fahrenheit. In a regular pot, the steam escapes and the temperature of the boiling water never gets higher than that. But high pressure raises the boiling point of water. That means that in a pressure cooker, the water becomes superheated, so it's boiling at a higher temperature than 100 degrees

Celsius—as much as 120 Celsius or 247 Fahrenheit."

Then, Jamal took over to explain how a pressure cooker lets people cook meats and other foods faster, but still keeps them moist because of the trapped steam. Finally, Pat explained their invention—a programmable pressure cooker.

"This automated program works with a sensor in the pressure cooker pot to keep the pressure and temperature at the right levels for the selected dish."

When Pat and Jamal finished, the class applauded. Then Paloma raised her hand.

"That was very interesting," she said, in her usual loud, clear voice. "I have a question. Where is the safety valve in your diagram?"

Jamal looked nervously at Pat. He also knew that Pat had trouble talking to Paloma. But somehow Pat stayed calm. He stepped up to the computer, hit a key, and brought back the first diagram.

"That's a good question...Paloma," he said with just a hint of nervousness. "Here's our safety valve. Without a way of letting off steam, the pot would explode." Suddenly Pat looked very nervous again. Then he tried another joke. "Uh, I guess everyone needs a way of letting off steam."

Paloma laughed and so did a few other kids. Pat looked very relieved.

"Okay, Pat and Jamal," Ms. Taylor said. "Thank you for a very interesting report. That looks like a good possibility for our class presentation at the fair. Now, we have one more, let's see...oh, yes! Stanley Roberts."

Stanley had been sitting impatiently and he jumped out of his seat near the front of the room. He was wearing his best business clothes, trying to look the part of CEO of StanTastic Industries. He had on a blue blazer, tan chinos, a white shirt, and a red-and-blue-striped tie. His laptop was already open on Ms. Taylor's desk, and he pulled a slim remote control out of his blazer pocket and hit a button. A video began to play on the screen. It was a video of Stanley standing in front of a hand-lettered sign that read: STANTASTIC INDUSTRIES.

"Hi," Stanley said in the video. "I'm Stanley Roberts, President and CEO of StanTastic Industries. Today I want to tell you about an amazing invention that will not only revolutionize cooking, but will save so much energy that the world will cut its fossil fuel use and global warming will be stopped."

Einstein could hear Paloma mutter in her seat. "Gee, don't be so modest."

"Once my new invention is introduced at the Sparta Middle School Spring Fair, people will be lining up in the stores to buy my incredible energy saver."

On the screen, Stanley held up a normal-looking pot. "This is the StanTastic Stupendous Low-Cook Pot. A new *low-pressure* cooker can cook food with much less heat than a normal pot."

A diagram appeared on the screen. In the video, Stanley continued, "Our vacuum pump lowers the pressure in the pot. This makes it possible to boil water at much lower temperatures. So water comes to a boil faster and with less energy. With the StanTastic Stupendous Low-Cook Pot, people will be able to cook food much more quickly and therefore more cheaply."

An image of melting icebergs appeared on the screen. Stanley's voice could be heard in the background.

"This is the choice: You can keep using more energy, burning fossil fuels, and increasing global warming. You can cook the old-fashioned way and destroy the earth's climate. Or, you can start using the StanTastic Stupendous Low-Cook Pot, save energy, and save the planet. The choice is yours!"

Dramatic music played from the computer's speakers as the video ended. Most of the kids in the class applauded. A few cheered. Einstein noticed that even Pat and Jamal applauded.

"I think we have a clear winner," Ms. Taylor said, smiling. "Stanley, you've done a great job. With this invention, I think this class just might win the science prize at the fair."

Stanley was beaming with happiness. He actually bowed.

"Thank you Ms. Taylor. I won't let the class down. We're going to beat the seventh and eighth graders like we beat them in the snow sculpture contest."

Paloma raised her hand.

"Uh, Ms. Taylor, are you sure that the science is right in that invention?"

Ms. Taylor didn't say anything for a moment. Then she cleared her throat. "Paloma, that's an excellent question, but I think the class should answer it."

A few of the kids groaned in protest.

"No, this will be a good test of your science knowledge," Ms. Taylor continued. "And what's at stake is our chance to win the science prize."

Stanley raised his hand and waved it until Ms. Taylor couldn't ignore him.

"Yes, Stanley?"

"Ms. Taylor, everyone could see I made a great presentation," he argued. "Just because Paloma and Einstein didn't think of it, doesn't mean it's wrong."

Einstein raised his hand.

"Yes, Adam?" Ms. Taylor said.

"Ms. Taylor," he replied. "If we put that invention up as our presentation, not only will we lose the competition, but we'll be the laughing stock of the school. Stanley's low-pressure pot won't work, and I can prove it."

Can you solve the mystery? Why won't Stanley's low-pressure pot cook foods faster and save energy?

"Oh, come on, *Einstein,*" Stanley complained. "I *know* there's nothing wrong with my science. Under low pressure, water will boil faster and with less energy."

"That's true," Einstein nodded. "Water boils more quickly when the air pressure is lower, but that doesn't cook foods any more quickly. In fact, it's exactly the reverse. On top of a high mountain, where the atmosphere is thinner and the air pressure is lower, you can't even cook a hard-boiled egg in a pot of boiling water."

"You can't?" Stanley asked, surprised. Then he caught himself. "What do you mean, you can't?" he demanded.

"Yes, under low pressure, water will boil at a *lower temperature*. But the thing you're missing is that—the water will be at a lower temperature. When you cook, the important thing is that the water is hot enough to cook your food. When you cook a hard-boiled egg, you don't really care that the water is turning into steam. You care that the water is hot. But if the water is boiling at a lower temperature it will be cooler, so it won't be able to cook your food as well."

"Uh, I..." Stanley stammered and his face turned red. He stared at the remote control in his hand. He looked very upset. "It's not fair!" he shouted. "I made the best presentation."

"Yes, you did," Ms. Taylor said gently. "But it's a science contest, so we have to make sure the science is correct. I think that Jamal and Pat's project should represent the class."

A dejected Stanley started to return to his seat. In spite of himself, Einstein couldn't help feeling sorry for him. He raised his hand again.

"Yes, Adam?" Ms. Taylor asked.

"I have a suggestion," Einstein began. "Since Stanley did make the best presentation, why doesn't he . . ."

Paloma interrupted him. "Why doesn't he make a video for Pat and Jamal's invention! Great idea, Einstein!"

Ms. Taylor laughed. "Yes, that is a good idea, *Einstein*. Stanley, would you like that?"

Stanley's face had a broad smile on it. "I'll have my advertising department get right on it," he said.

The class laughed and Paloma turned around in her seat to face Einstein. "He never stops, does he," she said.

"That's Stanley," Einstein agreed. "I just hope he doesn't eat his assignment."

For once Paloma didn't spot Einstein's joke coming. "Eat it?" she asked. "Why would he eat it?"

"Because," Einstein told her with a big grin, "he thinks it's a piece of cake!"

From: Einstein Anderson

To: Science Geeks

Experiment: How Dense Can You Get? An Ice Melting Race

Stanley's latest great invention, the low-pressure cooker, got me thinking about water and all the amazing things it does. Water is one of the most common substances in our environment. Seventy percent of the Earth's surface is covered with water. It's also a big part of ourselves—the adult human body is about 60% water and babies' bodies are as much as 78% water! It's something we see and use every day and something that animals and plants cannot live without. Water breaks things down—think of the Grand Canyon, which was made by water cutting through rocks. And it builds things up: our bones are more than 30% water.

So let's take a deeper dive into the chemistry of water and how it is affected by temperature. How about an Ice Melting Race?

Here is what you need:

- Water
- A scale
- Measuring cup
- Salt
- Food coloring
- Ice trays and a freezer
- A drinking glass
- A grease pencil or marker that will work on glass
- A timer
- 2 glass (or clear plastic) bowls
- A coin to flip
- A friend

You will want to make the ice the day before the big race. Fill your measuring cup with water and add a few drops of food coloring, then pour the water into an ice tray and put it in the freezer. Do the same with another measuring cup full of water, but use a different food coloring color so you will end up with two trays of ice in two different colors. If you need to add some plain water to fill up the ice tray, go ahead.

At the same time, fill a drinking glass about half full of water. Weigh the glass with the water in it and write down the weight. Use a grease pencil or marker to mark the water level and put the glass in the freezer with your ice trays overnight.

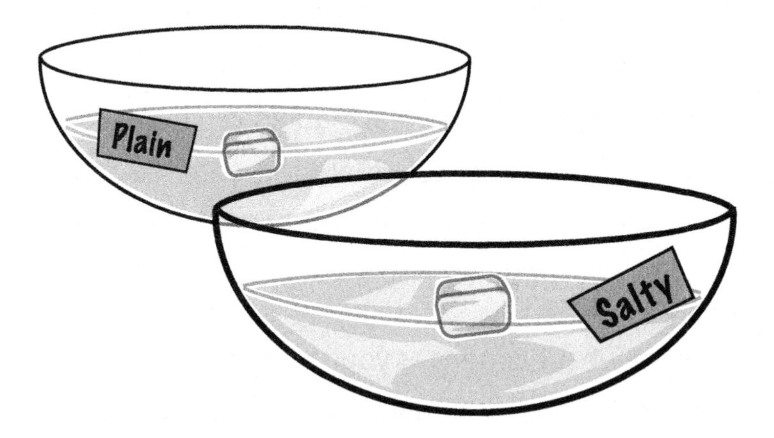

The next day, after the ice is frozen, fill your 2 bowls about ¾ full of water. Leave the water plain in one bowl but add salt to the other bowl and stir until the salt dissolves. Put in as much salt as you can until it will no longer dissolve.

Take your ice trays and the drinking glass out of the freezer. Look at the water level in the drinking glass.

When the water froze it expanded, so the top of the ice is now above the line you marked when you put it in the freezer. Weigh the glass again. Note that the glass weighs the same amount with ice as it did with water. Leave the glass of ice on the table while you have your race.

Now it's time to start the race. Flip a coin to decide who gets the plain water and who gets the salty water. You and your friend are each going to choose one of the colors and put ice cubes into the water at the same time. Set the timer and watch. Whichever ice cube melts first is the winner!

Observe what happens as your colored ice cubes begin to melt. Do both ice cubes float? Look closely at the different ice cubes. Do they float at the same level? Does one stick up higher out of the water than the other one? Do both ice cubes melt the same way? Which cube melts faster? If you have a camera handy, take a snapshot of the bowls as the cubes melt or draw what you see. Declare the winner!

The Science Solution

First of all, take a look at the glass where you froze the water. Now that the ice in the glass has melted, what do you notice? The water line is back to where it was when you started. We have shown that when water is frozen into ice, it expands but its weight does not change.

So the first question is, why does ice float? It was an excitable Greek scientist called Archimedes who first explained why things float about 2,200 years ago. He got into a full bathtub and noticed that water sloshed over the side. When he realized he could float because the weight of his body took up more space (volume) than the water it displaced (his body was less dense than water), he supposedly jumped out of the tub and ran naked through the streets of Athens, shouting "Eureka!" which means, "I found it!" in Greek.

I wouldn't try that if I were you.

So, since frozen water takes up more space (volume) than the same weight of liquid water, an ice cube (or an iceberg) will float. Water is more dense than ice.

And what about the ice melting race? Why did the ice in plain water melt so much faster than the ice in salt water? The answer lies in the patterns you saw when the colored ice cubes were melting.

The cube in plain water melted by a process called "conduction and convection." Heat energy travels from warm areas to colder areas. Since water is warmer than ice, the warmer water conducted heat to the ice and melted it. When you saw the colored water falling to the bottom of your bowl of fresh water, you were seeing convection in action. The cold water from the ice cube is more dense than the warm water, so it fell to the bottom of the bowl. That left the remaining ice in contact with the warm water, so the process of conduction and convection continued—the warm water kept melting ice and the newly melted cold water kept falling to the bottom of the bowl until the whole ice cube was melted.

Why didn't this happen in the bowl of salt water? When you added salt to the water you made a mixture that is more dense than plain water. The molecules of salty water are packed more closely together than the same volume of plain water and it weighs more. That is why the ice cube floated higher on salt water than on fresh water (ice is less dense than plain water and much less dense than salt water). It also explains why you did not see the colored water fall to the bottom in the bowl of salt water. Dense salt water does not cause convection currents like plain water because the salt water is more dense than the just-melted–ice water. Since the ice water did not sink away from the cube in the salty water, it took longer for the cube in salt water to melt.

Incidentally—back to Stanley's pressure cooker—the boiling and freezing temperatures of water are affected by both air pressure and salt content. While most people know that water boils at 100 °C (212 °F), this is at sea level. Water boils at just 68 °C (154 °F) on the top of Mount Everest, and deep in the ocean near geothermal vents, where the pressure is very high, water can remain in liquid form at temperatures much higher than 100 °C (212 °F).

If it has salt in it, water boils at a higher temperature, so salty water will cook your spaghetti faster.

Eureka!

The Untouchable Toes

The Sparta Middle School Spring Fair was a big event. The students looked forward to it all year. Parents and folks from the community flocked to the fair. In addition to the science competition, there was a bake sale, a used book sale, and a bunch of old-fashioned carnival games—like bobbing for apples. All of the money raised helped pay for the big year-end school trip to the state capital.

Each class was supposed to come up with a booth to help raise money. Since Einstein had done such a good job of settling the argument over the science project, Ms. Taylor had put him in charge of the class booth for the fair. He and Paloma had worked on it for several days. They'd gotten a bunch of the other kids in class to help out. Even Stanley was part of the team.

"Remember," Einstein texted everyone the day before. "Our booth is SECRET! Tell no one!"

The day of the fair was warm and sunny. The fair was held on the school playing field. Kids from every class got there early to set up their booths and put out their science projects. There were tables with cookies and cupcakes. A small stage was set up at one end of the field, and the band members were assembling their music stands and unpacking their instruments..

In the science area, Jamal, Pat, and Stanley set out the model of their pressure cooker, with colorful diagrams as a backdrop. The video that Stanley had made to explain it played on a monitor and, of course, Stanley was the star of the video. It was very dramatic, and everyone in the class thought he had done a good job. Meanwhile, Einstein, Paloma, and some of their classmates put up their class booth. They had it almost finished when a nasty voice called out.

"Impossible tricks? What sort of stupid booth is that?"

Einstein knew who it was without looking. It was Gary, the eighth grader who had been so

MOVE YOUR FOOT IN A CIRCLE
GIVE IT

angry when the sixth graders won the snow sculpture competition. Here he was again, walking toward them. Next to him was Andrea, another eighth-grade kid.

Gary was a big, burly blond kid with the thick neck of a football player. His biceps bulged under the sleeves of his Sparta Middle School T-shirt. Andrea was tall, too. Her hair stood up in a thick, curly mop around her head.

Gary came right up to Einstein and pointed to the sign above the booth. It read, IMPOSSIBLE TRICKS. WE DARE YOU TO TRY ONE.

"What's the point of trying something impossible?" he asked with a sneer.

"Yeah," said Andrea, as she towered over Paloma. "What's the point of that?"

"Well," Einstein replied calmly, as he pushed his glasses up onto the bridge of his nose, "The point is, we say it's impossible, but maybe you think it isn't. So you buy a ticket and spin the wheel. Then if you can do the impossible trick, you win a prize."

Gary stared at the big wheel they had set up in their booth. Like a wheel on a game show, it was divided into slices like a pie. On each slice was written an impossible trick. One pie slice read, LIFT YOUR LEG. Another read, MOVE YOUR FOOT IN A CIRCLE. A third read, STAND AGAINST A WALL AND TOUCH YOUR TOES.

Gary read each one carefully. "Those are stupid!" he declared when he'd finished. "Those aren't tricks at all. Our booth is a lot more fun."

"What's your booth?" Paloma asked.

"A fortune teller," Andrea said with a mean laugh. "Now, *that's* fun. Everyone will want to buy a ticket to our booth. We'll raise a lot more money than you."

"Maybe," Einstein said. "But I think once people try one trick, they're going to want to try them all."

"No way," Gary muttered. "Come on, Andrea. Let's leave these losers."

"I guess you're scared to try," Paloma said as the two eighth graders began to walk away. That stopped them in their tracks. They turned around angrily.

"Scared? Of you?" Andrea cried.

"I tell you what," Einstein told them, calmly. "We'll let you spin the wheel and do one trick for free. If you do it, you get a prize."

"For free?" Gary asked, looking doubtful. But Einstein nodded his head.

"Just try it," Einstein encouraged him. "You won't be sorry."

Gary shrugged his shoulders and gave the wheel a spin. It went around and around and finally stopped with the arrow pointing to the words, STAND AGAINST A WALL AND TOUCH YOUR TOES.

Gary laughed loudly. "Touch my toes?" he crowed. "That's the stupidest thing I ever heard. You think *that's* impossible?"

"Just try it," Paloma said. "But you can't bend your knees."

"Sure, I know," Gary said.

Paloma pointed to the wall of the school building. "There's the spot," she said. "You have to begin with your back and your heels touching a wall. Your feet have to remain against the wall as you bend."

"So what?" Gary asked as he walked to the wall. "I can touch my toes anywhere."

"Sorry," said Einstein, "but it can't be done."

Can you solve the mystery? How does Einstein know that Gary cannot touch his toes without bending his knees when his feet are against a wall?

"What a joke!" Gary said as he placed his back and his heels against the building's outer wall. "I can't figure out why they call you Einstein, if you think this is impossible."

He started to bend over, but almost at once he lost his balance. Gary looked around angrily, as if someone had pushed him. But no one was standing near him.

"Let me try that again," he said. Carefully he put his heels and back against the wall, then slowly tried to bend over. This time he nearly fell down.

"There's something wrong with this wall!" he declared, looking at the bricks suspiciously. He moved a few feet and tried it again and then again. Finally his friend Andrea burst out laughing.

"They're right!" she giggled. "It's impossible."

Gary looked at Einstein angrily. "That's right, impossible. No one can do it."

"That's just what we said," Einstein nodded.

"Oh, yeah?" Gary grumbled. Then he added, sounding a little curious, "Uh, how come? I mean I can touch my toes usually."

"It's standing against the wall that makes it impossible," Paloma explained. "It's all a matter of your center of gravity. It's like the balancing point in your body. When you walk or even when you're just standing still, you keep shifting slightly to make sure your center of gravity is over your feet. You do it without thinking about it."

"I do?" Gary asked. Now he seemed genuinely interested.

"Sure," Einstein told him. "You learn to do it when you're a baby and then it's automatic. But if your center of gravity isn't over your feet, then you fall over. It's just physics."

Paloma continued, "When you bend over, most of your body is in front of your feet, so you have to lean back a little to balance. But with a wall at your back, you can't lean backwards. So your center of gravity is in front of your feet and you fall over when you try to touch your toes. It's like a football player making a diving catch. Once his weight is too far in front of his feet, he's going to fall down."

Gary stepped away from the wall and tried to touch his toes.

"Hey!" he said, sounding surprised. "I can do it now!"

"Let me try that!" Andrea said. She took a dime out of her pocket and placed it on the table in front of the booth. Then she pushed past Gary to stand next to the wall.

Paloma turned to Einstein and said quietly, "That was some trick Einstein."

Einstein looked surprised. "It's just physics, Paloma," he said. "You know that."

"I didn't mean that trick," she whispered. "I meant the trick of taming Gary and Andrea. You should be a lion tamer in the circus."

Einstein started to laugh. "Hey, that reminds me," he said. Paloma groaned. She knew what was coming next.

"What circus performers can fly in the dark?"

Paloma just frowned, but it didn't matter. Einstein was already telling her the punch line.

"Acro-bats!"

From: Einstein Anderson

To: Science Geeks

Experiment: Amazing Feats and Boys vs. Girls

Don't you just love it when science beats logic? I mean, who would think that Gary couldn't touch his toes, just because his back was to the wall? Want to amaze your friends with some more feats involving the center of gravity? Try this!

Can you balance a quarter on a dollar bill?

Here is what you need:

- A quarter
- A dollar bill (or any piece of paper money)

Hold the money upright on its side and try to balance the coin on the edge of the paper. No matter what you do, it will fall off. It seems impossible to balance a heavy coin on such a thin edge.

Now try this—fold the paper money in half, lengthwise, and crease it. Then, open the bill to a wide V and balance the quarter on the point of the V. Slowly, pull both ends of the dollar apart until the bill is straight. The quarter will now balance on the edge!

Here's another one:

Here is what you need:

- 2 or more boys
- 2 or more girls
- 1 chair you can lift

Turn the chair sideways next to a wall. Ask a boy to stand with his feet just in front of (not under) the chair, and bend to a 90° angle with his head on the wall. Have him lift the chair and then try to stand up. Now have a girl do exactly the same thing—head on the wall, feet in front of the chair, lift the chair and stand up. Try again with another boy and another girl. What do you notice?

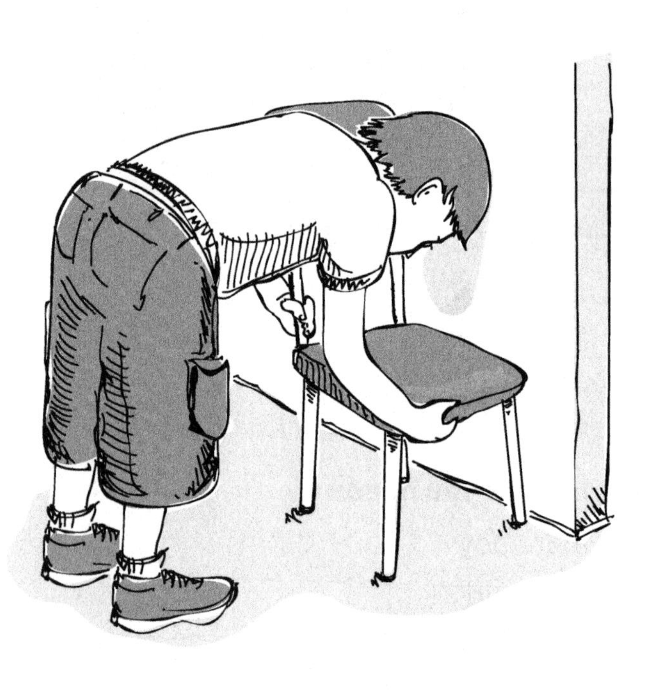

The Science Solution

These tricks—including the fact that Gary could not touch his toes with his back and heels against the wall—all depend on the "center of gravity," or center of mass, of the person or object involved. The center of gravity is the place in a system or body where the weight is evenly dispersed and all sides are in balance.

Since the force of gravity acts on everything on Earth, finding the center of gravity is an important part of building a bridge, or carrying something, or doing gymnastics. Have you ever played "limbo" at a party? It's that game where two kids hold a stick over the ground and others try to go under it, leaning backwards, without falling down. The reason why you finally fall is usually because your legs are not strong enough to hold you. But even a person with the strongest legs in the world would eventually fall down if they couldn't keep their feet under their center of gravity.

If you are still wondering why girls can stand up with the chair much more easily than boys, it's because the center of gravity in a girl's body is usually lower than the center of gravity in a boy's body. If you have trouble doing this in your classroom, try getting some older kids to volunteer. The boys may not like it too much, but the girls will be psyched!

Here's a weird idea—a horseshoe's center of gravity is outside the horseshoe. See if you can figure out where it is.

May the Force be with you!

CPSIA information can be obtained
at www.ICGtesting.com
Printed in the USA
LVOW04s0120251116
514317LV00008B/455/P